George Parris Copyright Claimed August 2019

The Neutrino

The Particle that Never Was

George E. Parris

George Parris   Copyright Claimed   August 2019

Summary

I am a chemist and only superficially schooled in physics. However, I became interested in the history of the development of physics in the period 1900-1950 a few years ago and as I read the stories of how various widely accepted ideas came into being, I was annoyed by the hubris of some of the personalities, dismayed at their politics, and entertained by their foibles. I was particularly drawn to the story of Wolfgang Pauli. In spite of the inclination of modern authors to venerate this Nobel Prize winner, my reading between the lines suggested that he was a man of mediocre talents disguised by a bombastic and overbearing personality. Naturally, his role in the *invention* of the "neutrino" came to my attention and fed my interest and skepticism.

As I understand the origins of the neutrino, it all circled around the fact that beta particles ejected from nuclei during the process:

$$n^o \rightarrow p^+ + e^-$$

were observed to have a continuum of kinetic energies rather than a discrete quantum-driven value as observed for other nuclear processes (typically producing electromagnetic radiation, not charged particles). The competition between Meitner's group and Rutherford's group (specifically Charles Drummond Ellis) to explain this observation was a battle of physics titans. And it seems to me that they were ultimately content to call it a draw. Then along came Pauli, very down on his luck and desperate to gain attention with nothing to lose, who was willing to solve the conundrum by postulating a mass-less particle that carries away kinetic energy.[1] Since no one had a reasonable argument against it, the idea persisted (festered?) until ultimately Enrico Fermi[2] dignified it by rationalizing it in a paper.

The Day the Neutrino Died: 3/1/2014[3]

Here is how I see the problem: The fundamental flaw in the thinking of everyone mentioned above was that *they all assumed the nucleus to be a point mass and point charge.* As such, there was no option for angular momentum of the nucleus and Gauss's laws do not come into play. In the discussion that follows, I claim to show that the continuum of beta radiation kinetic energy observed from collapse of a neutron to a proton can be explained *without invoking neutrinos*. Loss of some of the quantum of energy can be accounted for by changing the angular momentum of the nucleus and/or by the

---

[1] This is clearly a contradiction of terms that only an arrogant fool could propose and only people accustomed to seeing things they did not believe could accept.

[2] I think this is Fermi's biggest and most embarrassing mistake and that takes some doing.

[3] This is the day that I first did the calculations contained in this article.

phenomenon of "variable effective nuclear charge" (i.e., Gaussian surfaces within the nucleus).

But what about those observations of neutrinos? Yeah, what about them. As I read the various experiments, they usually fail until "adjustments" are made and eventually statistically significant results are obtained and then data collection stops or some new property (e.g., mass, variable mass, trans-luminal velocity, etc.) is declared discovered as a new property of the neutrino. Sorry folks, but to me this whole thing looks like a pile of shifting sand.

Regardless, I have sent these ideas to various physicists who have ignored them. But if what I propose is demonstrably wrong, I would be happy to be educated. And, let me offer this challenge, even if neutrinos (of some sort) exist, why don't we see the phenomenon associated with a non-point (i.e., three-dimensional) nucleus, which I describe here?

What follows is my rendition of history, topped off with my suggestions and a few fictional anecdotes that express my cynicism.

George E. Parris

Gaithersburg, MD

August 2019

This idea was original with me on 3/1/2014

George Parris — Copyright Claimed — August 2019

The following is copyright claimed, but may be used in academia with attribution:

Forward

Theoretical physics is not my area of expertise and I rarely worry about it. But I became interested in certain types of radioactive decay and came across what, to my untrained and un-indoctrinated eyes looks like a grandiose mistake. Looking at particle physics from the outside in, it has always seemed to me that the practitioners are rather cliquish with a lot of arrogance relative us mere mortals who prefer a physical model over an elegant equation filled with linear algebra and Greek letters. Yes, we mortals talk to God on a daily basis. But particle physicists think that God asks them for advice. I'm reminded of the discussion among participants at the 1927 Solvay Conference when Paul Dirac is reported to have said something like:

> *"I cannot understand why we idle discussing religion. If we are honest – and as scientists, honesty is our precise duty – we cannot help but admit that any religion is a pack of false statements, deprived of any real foundation. The very idea of God is a product of human imagination. [...] I do not recognize any religious myth, at least because they contradict one another. [...]"*

And Wolfgang Pauli quipped:

> *"Well, I'd say that also our friend Dirac has got a religion and the first commandment of this religion is 'God does not exist and Paul Dirac is his prophet'."*

As a purely practical matter, when you start thinking yourself to be a god or among gods or that there is no higher power than yourself, you are probably living in a world of euphoria and likely to overplay your hand. Of course, much of Dirac's criticism of religions is justified precisely because the founders of the religions had the same mind-set that the people at the Solvay Conference were displaying…complete faith in their superiority.

In general, I have been somewhat dismayed in the technique used by particle physicists who resolve many of their theoretical problems by inventing new particles or forces and giving them clever names. I would think that this would be the last option, not the first option. And, I think that physics acquired this habit in 1931. In this story, I display my ignorance of theoretical and particle physics by attacking the early life of Wolfgang Pauli, who I have grown <u>not</u> to admire. He was an arrogant youth running in the right circles primarily because of family connections. He hid his incompetence and immaturity by setting himself up as the critic and bullying the weaker personalities around him. As a scientist, I would certainly be happy to have those qualified in the art correct me and if you show me to be quite wrong, I'll gladly recant.

Wolfgang Pauli (1900-1958) is a respected theoretical physics and Nobel laureate (1945) for his contribution of the "Pauli Exclusion Principle." The exclusion principle, states that no two particles can have the same set of quantum numbers. The practical implication of this for chemists is that electrons have only four quantum numbers and the last one (which designates the magnetic moment of the electron) can only have two values. Thus, two and only two electrons can share the first three quantum numbers, which define a volume of space in which the electron has a high probability of being found.

The physicists reach this conclusion through baroque mathematics. The chemists look at it from a more practical standpoint: Their rationale is that the negatively charged electrons repel one another, but when they have opposite "spins" (i.e., magnetic moments, represented by arrows pointing up or down↑↓) they can occupy the same space. In chemistry texts, no one ever seems to explore this. I suspect they are intimidated by the physicists. (Chemists are generally not very good with math.) Silly me, it seems that the whole phenomena boils-down to the idea that electrons (paired with opposite magnetic moments interpreted as "spins") are basically balancing electrostatic repulsion with magnetic attraction (N attracting S and S attracting N). I fail to see the mystery here. Obviously, any attempt to put a third electron into the same space (already occupied by two negatively charged electrons) will be repelled since the third electron must be attracted magnetically by one of the electron pair and repelled magnetically by the other. What's the big deal? It's all just an application of electrostatics and magnetics determining the disposition of electrons in atoms and ions and explaining covalent bonds at least qualitatively. I'll leave it to the theoretical physicists to crank out the quantitative details. Chemists are more likely to take an empirical approach.

Moreover, given that electrons sometimes behave as particles and sometimes behave as waves, I have no problem dissociating the concept of "spin" from a physical object. I would guess that if there is actually a spinning electron particle, that spin would be conserved in the direction of rotation of the quantum of electromagnetic radiation with which the particulate electron is in equilibrium.[4] Let me suggest an experiment: start with an electron that is behaving as a particle, send it though an environment where it is diffracted as a wave and see if it still has the same "spin" when it reaches the other side? Also check to see if the electron (+/-) acting as an electromagnetic wave is polarized (+/-) and if the polarization correlates with the spin of the electron behaving as a particle. Surely, some of you physicists

---

[4] The interconversion of matter and energy (recall $E \leftrightarrows mC^2$) must conserve energy as mass (i.e., the limit of the kinetic energy as velocity approaches zero), and the electric and magnetic vectors of the photon as charge and magnetic moment, respectively.

have done this experiment (e.g., the Faraday Effect of an electron wave).

In any event, Pauli seems to me to have been a very lucky young man: He was in the right place at the right time with the right connections to gain recognition. And that is where, in my opinion, the big mistake happened. Let's see what Pauli is up to. But, first let's look at some of the work leading up to the great error.

There are a number of useful descriptions of this period in scientific history. I found Ruth Lewin Sime's book *Lisa Meitner: A life in physics* (1996 University of California Press, 526 pp.) to be especially useful.

Chapter 1. Electron Spin

The 1800s

Johann Geissler was a clever glassblower and produced a variety of tools in the 1850s that both entertained the masses and intrigued the scientists who followed in the footsteps of Gilbert, von Guericke, Du Fay, van Musschenbroek, Franklin, Priestley, Galvani and Volta. V.S.M. vander Williben developed Geissler's toy into the cathode ray tube by 1855. Michael Faraday (1791-1867) overcame poverty and personal disabilities to gain a position in Humphry Davy's laboratory. Experiments with moving conductors and magnets linked magnetism and electricity, which had previously been viewed as separate phenomena, and led directly to the electric generator and electric motor. He also linked electricity and magnetism to light (electromagnetic radiation) with the Faraday Effect (rotation of the plane of polarized light in a magnetic field). The realization that electricity, magnetism and light were inseparably comingled inspired James Clerk Maxwell (1831-1879), who had developed an interest in color, to develop concepts of lines of force and explain light in terms of electromagnetic vectors (1864). He even made an accurate estimate of the speed of light from his electric data. Looking at it another way, the fact that Maxwell's equations could only be mutually consistent if electromagnetic waves moved at a speed identical to that independently measured for visible light linked light to electromagnetic radiation.

By 1859, Kirchhoff had shown that the discharge of a cathode ray tube was unique for each element. From the standpoint of atomic theory, George Johnstone Stoney (1826-1911) is important and forgotten. He was the first to assume that there was a basic unit of electricity (1874) and called it an electron (1891). In the meantime (1881), J.J. Thomson (1856-1940) and J.J. Balmer (1825-1898) began experimenting with cathode ray tubes. In 1885 Balmer was working out the rules that predicted the observed line spectra of hydrogen atoms. These are caused by electronic transitions between energy states (orbitals). Lorentz and P. Zeeman studied the spectra of atoms very carefully and examined the effects of magnetic fields on them. Sodium is convenient for study because it is a relatively volatile metal and produces a reasonably simple atomic spectrum dominated by a bright yellow line at 589 nm. Close examination shows this line to be a doublet with two peaks 0.6 nm (0.0021 eV) apart in the absence of a magnetic field. In the presence of a magnetic field, Zeeman observed additional splitting in 1896. Concurrently, Thomson used Crookes's highly evacuated tube to study the nature of the particles that were causing all the excitement. In 1897, Thomson identified the electron as the fundamental unit of charge and measured the ratio of its mass to charge (m/e) based on the curvature of its pathway in a magnetic field.[5]

The 1900s

Subsequently, Lorentz and Zeeman were able to confirm m/e of the electron from the magnitude of the splitting in a field of known strength. However, the simple electric model of the electron (charge and mass) did not explain the "anomalous Zeeman effect," i.e., extra (unexpected) splitting of some spectral lines in a magnetic field. This was not resolved until 1925.

---

[5] Interestingly, the particles were not affected by electric fields. It turned out that the CRTs then in use still contained enough air to ionized and neutralize the external field.

Often forgotten in the story that follows is the remarkable work of Arthur Compton (1892-1962). Although he is most noted for his leadership in the Manhattan Project in the 1940s, he did some impressive fundamental science before 1920. In particular, he envisioned the electron as having a magnetic moment derived from rotation (spin) on its axis and published this in 1921 without attracting much attention [A.H. Compton, The Magnetic Electron, *J. Franklin Inst.* 192, 144 (1921)]. This conclusion followed primarily from his analysis of bulk magnetic properties of crystals (not theoretical calculations of quantum mechanics). At the time, Compton was an unknown scientist publishing from a little-known university (Washington University of St. Louis). But, the hypothesis of a spinning electron must go back even farther to A. L. Parson [Magneton Theory of the Structure of the Atom, *Smithsonian Misc. Collection*, 1916, 80 pp.] as cited by Compton:

> *"...many of the magnetic properties of matter receive a satisfactory explanation on Parson's hypothesis, that the electron is a continuous ring of negative electricity spinning rapidly about an axis perpendicular to its plane, ... While retaining Parson's view of a magnetic electron of comparatively large size, we may suppose with Nicholson that instead of being a ring of electricity, the electron has a more isotropic form ..."*

> *"May I then conclude that the electron itself, spinning like a tiny gyroscope, is probably the ultimate magnetic particle."*

However, within a couple of years Compton's work on x-ray scattering would bring him the Nobel Prize (1927) and prove that Max Planck was correct about the quanta of electromagnetic radiation.

Ironically, about the same time, Einstein and Wander Johannes de Haas were also looking at gross magnetics properties of matter that hinted at intrinsic electron magnetism [A. Einstein, W. J. de Haas, 1915, Experimental Proof of Ampère's Molecular Currents, *Deutsche Physikalische Gesellschaft*, 17: 152-170]. This work on the Einstein-De Hass effect was interrupted by World War I.

Scientists have a saying, "Theories may come and go, but good data will last forever." And, so it was with the work of Otto Stern and Walther Gerlach. Text books today greatly simplify this episode in quantum physics with the clarity of 20/20 hindsight. But the story is fairly complicated. By 1921, the notion of quantitation of energy levels was starting to take hold. Max Planck had postulated the quanta of energy associated with electromagnetic radiation (1900). Einstein had proposed quantized energy levels from the photo electron effect (1905), which was confirmed by R.A. Millikan in 1916. The work of Balmer had been extended by Ryberg (1888) and captured by Bohr (1913) in a model of the atom; and

G.N. Lewis was beginning to use all this information to explain chemical bonding and reactions in 1916. But, all of this had to do with the potential energy of electrons in an electric field; nothing had yet been done about magnetic quantization, i.e., the orientation of atoms in space relative to an axis defined by a magnetic field (i.e., space quantization).

Based on the Bohr model, an electron was assumed to circle the nucleus of an atom. This movement of a charged particle should cause an orbital magnetic moment (like any other circuit). Based on rough calculations, Stern estimated the power of an inhomogeneous magnetic field needed to detect this induced magnetic moment and Gerlach concluded that he could deliver such an instrument. If classical mechanics applied, neutral silver atoms passed through the magnetic field would be oriented in many different directions and they would be displaced in a continuous smear, but if the silver atoms were confined to specific quanta of magnetic energy (as suggested by the Zeeman effect) they would be sorted into distinct groups. After trial runs in which the instrument was refined to provide greater sensitivity, Stern and Gerlach showed that, indeed, there were preferred orientations of the atoms in the magnetic field. To this extent, the idea that "space" (i.e., magnetic orientation) was quantized was confirmed and classical ideas were put on the shelf. Quantum mechanics seemed to rule all electromagnetic dimensions. But, was the orbital magnetic moment the cause of the observation?

Enter Pauli

The brash young physicists Wolfgang Pauli (then only 21 years old) who was a firm believer in all things quantum wrote to Gerlach (February 17, 1922), "*Hopefully now even the incredulous Stern will be convinced about directional quantization.*" I think it is typical of Pauli to celebrate his "victory" like an arrogant school boy while totally missing the point of doing careful research to test a hypothesis.

Ironically, the simple model of quantum mechanics that Pauli understood did not explain the magnitude of the effect. Further refinements of the quantum theory also failed to resolve the issue and even created more qualitative problems. It was not until George Uhlenbeck and Samuel Goudsmit popularized the "hypothesis of the magnetic electron" among the European quantum elite in 1925 that the Stern-Gerlach and the anomalous Zeeman Effect fell into place. Goudsmit had worked with Zeeman at the University of Leyden where the study of electricity went all the way back to the Leyden jar of van Musschenbroek (1746). The young students published two papers and they apologetically cited Compton in the second paper.

This hypothesis was not as easily digested by the European theoretical quantum physicists. In describing the exclusion principle, Pauli (1924) had created a requirement for a fourth quantum number. The first three quantum numbers define: (n) the energy of the electron, which is associated with its distance from the nucleus (i.e., radius of its Bohr orbit), (l) the shape of the orbital, and (m_l) the orientation of that orbital relative to an outside magnetic field. If no more than two electrons were to occupy the same orbital (something that the chemists, e.g., G.N. Lewis, saw happening) then a fourth quantum number was needed. But Pauli was probably not sensitive to electron-pair theory and viewed a fourth or more quantum numbers as mathematical options, not necessities. It is worth explaining that the first three quantum numbers (n, l, and m_l) all were related to the position of the electron (relative to the nucleus) in three-dimensional space and they were associated with the degrees of freedom of a particle in space. Thus, a fourth quantum number was hard to contemplate as a motion of the electron in three-dimensional space (e.g., defined in polar coordinates). Uhlenbeck rationalized that it must relate to the "spin" of the electron (i.e., an internal motion) and that a spinning electron would produce a magnetic moment (up or down in a magnetic field). Ironically, one of Pauli's post-doctoral associates, Ralf de Laer Kronig (1904-1995), in an attempt to put physical meaning to Pauli's fourth quantum number, suggested to Pauli in 1925 that the electron might have a magnetic moment; but Pauli so rudely condemned the idea saying "*...it is indeed very clever but of course has nothing to do with reality*" that Kronig was bullied into not publishing and apparently he dropped the idea altogether. Nonetheless, Pauli soon published his own Nobel Prize-winning paper [W. Pauli, "On the Connection between the Completion of Electron Groups in an Atom with the Complex Structure of Spectra", 1925, *Z. Physik* 31:765ff] and some people argue that Kronig had sparked Pauli's thinking about the exclusion principle.

Thus, Pauli's new quantum number had a physical meaning and 'm_s' was born. The hypothesis was very powerful in explaining empirical observations. Bichowski and Urey (1926) confirmed the general idea, but there were problems in the details.

From the stand point of the European quantum elite, the fact that the numerical calculation provided by Uhlenbeck and Goudsmit was off by a factor of 2 quantitatively raised great suspicion of the proposal coming from graduate students. Under attack by Pauli and Heisenberg, Uhlenbeck and Goudsmit began to "twist in the wind" [as if hung by the neck]. Pauli characterized their hypothesis as "false doctrine." Modern-day supporters of Pauli (see *The Joy of Quantum Physics* by Michael A. Morrison chapter 6) argue that Pauli's only objection concerned the physical model of a spinning electron. Their view (which has much merit) is that the electron magnetic moment does not need to be (and cannot be) explained by a classical model...it is what it is. But Pauli's attacks seem personal and not scientifically constructive. Rather than describing electron spin as "false doctrine," Pauli might have merely and more clearly stated that it was an unnecessary model of a clearly existent property (electron angular momentum). But, he did not. Indeed, Compton had already addressed some of the issues raised by Pauli in his 1921 paper.

Then, along came Llewellyn Hilleth Thomas, a math whiz who had studied under Eddington at Cambridge and thought that the factor-of-2 problem might be associated with relativistic effects between the nucleus and the electron. Bohr, who introduced him to the spin problem in 1925 in Copenhagen, and others were skeptical that relativistic considerations were very important. Nonetheless, Thomas soon published a paper (1927) that resolved the discrepancy and won over Bohr and Heisenberg. Pauli still refused to accept the phenomenon, but he conceded in a letter to Goudsmit (13 March 1926) that Thomas's correction to the intrinsic angular momentum of the electron explained the atomic spectrum fine structure. Thus, the Stern-Gerlach experiment was ultimately interpreted in terms of electron angular momentum, rather than orbital angular momentum.

For my part, I also think that the physical manifestation of the electron as a spinning sphere of defined radius is unlikely and is only consistent with the notion of continuous corporal existence of the electron in a defined

Bohr-like orbit. When we dissolve this notion and accept orbitals as defined spaces in which electron density exists, the dual nature of an electron as a particle and a wave becomes important. I see the electron angular momentum as a result of conservation of the magnetic vector of electromagnetic radiation; the charge on the electron as being a conservation of the electric vector; and the mass is the conservation of its momentum. Thus, when the electron behaves as a particle it must have these vectors (mass, charge and angular momentum) independent of physical description.

Volume does not seem to be a conserved quantity. Thus, when Compton (1921) calculates a radius of 10^{-12}m for the electron and notes that electrons interact with and scatter x-rays and gamma-rays (which have much different wavelengths), I am inclined to believe that the effective size of the electron does not matter. I would be inclined to set an upper limit on the *effective* size of an electron by the onset of Compton scattering (0.1 MeV) and a lower limit by the wavelength of light that initiates pair-production (0.5 MeV), i.e., 10^{-11} to 10^{-12} m. And, for practical purposes, this is the size of the *orbital* in which the electron (or electron pair) can be localized on a molecule. However, for a metal, the orbitals (electron bands) may reach macroscopic dimensions.

Chapter 2. Pauli: The Man with the Right Connections at the Right Place and the Right Time

Remember Pauli was born in 1900. What elevated him to the position that people cared what he thought by 1921? First, he had the correct pedigree. Pauli was from a well-known family that had bridged both the Jewish and Roman Catholic upper classes in Austria. The family ties included Ernst Mach (1838-1916) who was appointed to be his godfather (in spite of his advanced age) and mentored Pauli early in his career. Interestingly, Mach was one of the last hold outs in the belief in the existence of atoms. Pauli developed an early interest in the work of Albert Einstein on relativity and began a series of papers on relativity that established his reputation. However, his expertise seemed to be in articulating the topic so that others could understand it more than making any original contribution.

In 1918, Pauli was given a position in Munich to work for Arnold Sommerfield (1868-1951) who groomed many future Nobel Prize winners. At that time Bohr and Sommerfield were the world leaders in quantum physics of the atom. Whereas Bohr had envisioned circular orbits of electrons (requiring only one quantum number, n), Sommerfield anticipated circular and elliptical orbits and introduced the second quantum number (l) to distinguish the two. Ultimately, a third quantum number (m_l) was added to designate the three-dimensional distribution of the electrons. Pauli continued his interest in relativity; and after publishing a couple of minor papers, Sommerfield suggested that he write a major review of the topic, which Pauli had followed almost since its inception. In contrast, Pauli's thesis was on the quantum theory of ionized molecular hydrogen (H_2^+, the simplest molecule consisting of two protons and one electron). No one seems to have been impressed; his theoretical results did not match experimental results.

In 1920, Pauli was joined in Sommerfield's lab by Werner Heisenberg (born 1901). Pauli was later described by Heisenberg as a night owl who hung out in bars and cafés until late in the evening and then worked alone through the night. This lifestyle annoyed Sommerfield because Pauli rarely showed up for class in the morning and generally was not around before noon. The product that most impressed everyone from his graduate days was his flattering review of Einstein's relativity theory that was published in 1921 in *Encyclopädie der mathematischen Wissenschaften*. Of course, Einstein read the article and wrote a glowing review...after all, Pauli was the first reviewer of Einstein's work and was glorifying Einstein at a time when Einstein was hoping to win the Nobel Prize... *"...for his services to Theoretical Physics, and especially for his discovery of the law of the photoelectric effect"*.

Remember, there is a substantial amount of politics associated with the Nobel Prize and Einstein would reasonably have won it in 1919 for the theory of relativity after Eddington prove his prediction about the bending of light by gravity to be correct; but anti-Semitism was on the rise and relativity's obscure mathematics made it difficult to overcome the criticism of skeptics (e.g., Ernst Gehrcke and Philipp Lenard). This is where Pauli's review was most welcomed. Einstein was very pleased with the review (from a loyal follower) and could (under the circumstances) not have been critical of the review. Thus, Einstein made it sound as though Pauli was the most brilliant student to ever pick up a pen. The review almost turned the tide for Einstein in 1921, but the Nobel committee decided to award no one the prize while they decided what to do. In the end, they awarded the 1921 prize to Einstein in 1922 for the photoelectric effect (which had actually been demonstrated by R.A. Millikan in 1916, not Einstein). Once Einstein had won the Nobel Prize, his resonating approval of Pauli's *review* sounded like a ringing endorsement of Pauli *himself*.

Pauli (1921) and Heisenberg (1922) both moved on to the University of Göttingen to work under Max Born, who had just returned to his alma mater. Heisenberg continued with Born until 1925 when they collaborated on the concept of "matrix mechanics" that was soon overshadowed by Erwin Schrodinger's "quantum mechanics." Pauli used the matrix method to calculate energies of the hydrogen atom that were consistent with those done by Bohr. Heisenberg (alone) won the 1933 Nobel Prize for matrix mechanics. This may have been influenced by the national politics of 1933 (Heisenberg was a Nazi sympathizer and headed Germany's nuclear program during WWII). But Heisenberg fully acknowledged that Born and others had played leading rolls in the development of matrix mechanics. In the meantime, Pauli followed Neils Bohr to Copenhagen and in 1925 published the "exclusion principle." It is relevant that his Nobel Prize for the exclusion principle was not awarded until 1945. To this point, there is nothing (save Einstein's self-serving praise of Pauli's review) to make anyone believe that Pauli was more than an average member of a well-known research group. The thing that set Pauli apart was his arrogance and ability to intimidate colleagues and junior staff as we saw above.

Chapter 3. The Nucleus

The cathode ray tube, which evolved from the Geissler tube in the 1850s, gave way to three important discovers in 1895, 1896 and 1897.

Rontgen, Becquerel and Curie

In November and December of 1895 Wilhelm Rontgen (1845-1923), discovered that cathode rays seemed to give rise to an unknown but highly penetrating ray (the x-ray), which he could detect by its production of phosphorescence on appropriate materials. He soon found that the rays could also be detected by photographic plates and was soon using the technique to take x-ray photographs on the bones in his wife's hands. The publication of Rontgen's work immediately inspired experimentation all over the world. The day after Rontgen's results were reported in England, Silvanus P. Thompson (a gifted lecturer of science) repeated the experiment and gave demonstrations to the Clinical Society of London.

In France, Henri Becquerel (1852-1908) was inspired to see if he could induce highly penetrating rays from phosphorescence of minerals that he had been interested in, including a salt of uranium (potassium uranyl sulfate). His first phosphorescence experiments (presented in February 1896) appeared to be successful as he produced penetrating rays from his uranium salt. But, by March 1896, he found that the uranium salt all by itself (without stimulation by light) would produce images on photographic plates. On April 8, 1896, Silvanus Thompson published a note in *The Electrical Engineer* (XXI (414):354) explaining how he determined that the source of the Rontgen rays was any target placed in the beam of cathode ray (whether it fluoresced or not). That volume also included an article by Thomas A. Edison entitled "Are Rontgen Rays due to Sound Waves." By May

1896, Becquerel was convinced that the rays came directly from the element uranium and were not associated with phosphorescence. The success of Becquerel can be viewed as pure luck. Sooner or later someone was going to place a mineral of uranium on a photographic plate and make the discovery he made. However, he did study the phenomenon in some depth.

Becquerel was at the French National Museum of Natural History (in Paris). Meanwhile nearby at the University of Paris (Sorbonne) Pierre Curie (1859-1906) had recently finished his doctoral on the effect of temperature on magnetism (under Gabriel Lippmann) and married Maria Salomea Skłodowska (1867-1934) in the summer of 1895. Maria Skłodowska was a woman with a driving passion for science. She had come to Paris from Poland because she was not able to get an advanced education in Eastern Europe. In Paris, she could barely support herself, but she managed to obtain a basic science education and intended to return to Poland to teach. When she found out that, in her native country, she would never get a job as a professional scientist, she returned to Paris and married Curie. Her subsequent research was more or less done on her own, independent of academia.

By the spring of 1896, the results of Roentgen and Becquerel were widely known and Marie began experimenting with uranium. She used Pierre's instruments to measure the current produced in the air as uranium produced penetrating rays. In 1897 she had a daughter (Irene). Behind the laboratory where Pierre taught, they took over a small building where Marie conducted her research, often recruiting her husband to help. She discovered that the uranium-containing minerals pitchblende and torbernite were both producing more radiation in the air than pure uranium salts. Thus, she deduced that these minerals must contain small amounts of much more active elements. This observation prompted her to attempt to isolate the active element(s). Marie's successes caused Pierre to drop his research and spend most of his time working on her projects.

Their first attempts to identify the new radioactive element began 12 April 1898. Starting with about 100 g of mineral, they soon discovered that the amount of the element(s) of interest was so small that they had to scale up their operation to tons of mineral in order to isolate visible quantities. Nonetheless, by July 1898 they reported the isolation of a new element (polonium), which she named for her native Poland. And, on 26 December 1898, they announced the isolation of radium and coined the term "radioactivity." In the next 4 years, the Curies published a number of papers about the isolation, reduction and properties of polonium and radium. They noted that radium seemed to produce a radioactive gas, which was called "radium emanation."

Thomson and Rutherford

While all of this was going on, J.J. Thomson (1856-1940) used the Crookes (1897) tube to demonstrate that cathode rays are actually electrons and that electrons are negatively charged and behaved as small particles for which Thomson determined the ratio of mass to charge (m/e).

Ernest Rutherford (1871-1937) conducted his own experiments on uranium and discovered (1898-1904) that two types of ionizing radiation (i.e., radiation that causes ionization in air as observed by the Curies): positively charge alpha particles (which turned out to be helium nuclei) and beta particles (which turned out to be electrons). In studying the nature of the different types of radiation from radium and thorium, Rutherford and Frederick Soddy (1877-1956) isolated two gases that proved to be unreactive to all types of reagents. It appeared that the gases must be related to the family of elements associated with argon (isolated from air in 1894 by Ramsay). In 1903, Ramsay and Soddy proved that one of the gases was helium; the other gas ("unidentified radium emanation" in the paper) turned out to be radon (atomic number 86), which Ramsey isolated and characterized in 1904. But, the question of the nature of alpha particles was still unanswered. The technique that Rutherford used (1908) to show that that alpha particles are helium nuclei is interesting. Working with Hans Geiger (1882-1945), they produced a very thin-walled glass tube that alpha particles (as well as beta and gamma) could pass through. Rutherford allowed alpha particles to pass into the evacuated tube for a period of time and then passed a spark through the tube. Helium was identified by its atomic spectrum. Meanwhile, Geiger used the tube to develop the Geiger counter for measuring radiation. Accurate measurement of the amount of radioactivity led to the idea of radioactive half-life for elements like radium (half-life 1601 years) and polonium (half-life 103 years).

In 1909, Rutherford noticed scattering of alpha nuclei from various solids and convinced Hans Geiger to have his student (Ernest Marsden) do an experiment that showed that alpha particles readily penetrated gold foil, but were occasionally scattered in a range of angles. From this Rutherford deduced that atoms had small dense positively-charged nuclei surrounded by clouds of negatively charged electrons (1911). A modern picture of the atom was beginning to take shape. Since some of the spontaneous radiation observed was positively charged and involved relatively massive particles (indeed particles identical to helium nuclei), it was assumed that most of the spontaneous radiation involved decay of unstable nuclei to more stable nuclei. In these processes, elements were 'transmuted' into new elements, as the ancient alchemists had hoped!

Soddy's careful work revealed that some transmutations produced elements with atomic masses that differed from those same elements that were isolated from nature. This led to the notion of isotopes (i.e., atoms of the same chemical elements with different atomic masses), which was confirmed by J.J. Thomson and F.W. Aston using a crude mass spectrometer to separate ^{20}Ne and ^{22}Ne in 1912.

J.J. Thomson had not disappeared after discovering the electron; he was a continuing force in atomic theory. Thomson had gone on to work on positive ions in the Crookes tube and had been one of the first to guess that alpha particles were helium nuclei. He deduced that H^+ (the proton) was the fundamental unit of positive charge (circa 1908) and began to speculate about the existence of a "neutral doublet" containing both H^+ and e^-. By 1920, Rutherford had adopted the concept of a "neutral doublet" (H^+/e^-) particle and was actively searching for it. Rutherford and James Chadwick (1891-1974) spent the next 12 years looking for what they assumed was an electronically neutral particle that had mass about the same as a proton, i.e., the neutron.

Chapter 4. Beta Decay

Electromagnetic Radiation

Before going too far into this discussion, it is relevant to remind you that photons (energy electromagnetic waves) have been arbitrarily subdivided into groups ranging from the very long radio waves, to microwaves, to infrared radiation, to visible light, to ultraviolet to x-rays and very short wavelength gamma rays. For practical purposes x-rays are usually of lower energy than gamma rays, but the distinction is often based not on energy, but on source. Gamma rays are generally assumed to come from very high energy processes in the nuclei of atoms, whereas x-rays are generated by collision of electrons (cathode rays) with inner electrons of atoms.

If you apply Einstein's famous equation,

$$E = mc^2,$$

and Planck's postulate

$$E = hc/\text{wavelength},$$

you find that there is not enough energy in the photons of electromagnetic radiation to produce even small subatomic particles, until very, very short wavelength gamma rays (1.022 MeV) begin producing pairs of electrons and positrons (pair-production).

By 1900, emissions from radioactive elements were classified into three types based on penetrating power: alpha, beta and gamma; and Becquerel showed that the beta particles had the same m/e and charge as electrons (determined by Thomson). It was, thus, easy to believe that beta-rays were the same as cathode-rays (i.e., electrons with kinetic energy). By 1913, Soddy understood that emission of a beta particle (an electron) produced elements with one-unit higher atomic number and emission of an alpha particle produced elements with two units lower atomic numbers. The characteristic of the chemical element thus lay with the nucleus, specifically the nuclear charge. The chemistry of an element is determined by its electron configuration and the electron configuration is determined by the number of electrons balancing the nuclear charge.

Not only did some elements give off radiation, they gave off heat. A gram of radium (Ra) yields about 1000 calories per year as it disintegrates to radon (Rn) and ultimately lead (Pb). Rutherford and Soddy estimated in 1903 that these nuclear changes were producing thousands and perhaps millions of times as much energy per unit of mass as is observed in chemical reactions. There was a lot of energy to be found in the pitchblende (uranium ore) of North Bohemia and other locations, if a method could be found to ferret it out.

H.G Wells

Marie Curie isolated pure radium in 1911. It decayed at a seemingly unalterable natural rate, with a half-life of 1620 years. This decay rate controlled (limited) the rate of nuclear energy release. This limitation made it impractical to use pure radium as a source of energy. Enter H.G. Wells, one of the fathers of good science fiction. In 1914, on the eve of the Great War in Europe, he conceived and published a work of fiction entitled *The World Set Free*. In this book, Wells envisioned his fictional character (Holsten) achieving the goal of accelerating the release of nuclear energy to useful rates in 1933:

THE NEW SOURCE OF ENERGY
[written in 1914 by H.G. Wells]
"The problem which was already being noted by such scientific men as Ramsay, Rutherford, and Soddy, in the very beginning of the twentieth century, the problem of inducing radio-activity in the heavier elements and so tapping the internal energy of atoms, was solved by a wonderful combination of induction, intuition, and luck by Holsten so soon as the year 1933."

...

[By 1913, radium C and E were known to be bismuth isotopes (*Chemical News* 107, 97-9 (1913)). The

> production of a heavy radioactive gas (radon) was known from radium.]
>
> ...
>
> "He [Holsten] set up atomic disintegration in a minute particle of bismuth; it exploded with great violence into a heavy gas of extreme radio-activity, which disintegrated in its turn in the course of seven days, and it was only after another year's work that he was able to show practically that the last result of this rapid release of energy was gold. But the thing was done—at the cost of a blistered chest and an injured finger, and from the moment when the invisible speck of bismuth flashed into riving and rending energy, Holsten knew that he had opened a way for mankind, however narrow and dark it might still be, to worlds of limitless power." [H.G. Wells never actually explained what catalyst was used to accelerate the natural radioactive decay.]

The important scientific idea captured by Wells was a rapid, and hopefully controllable, release of nuclear energy. The most interesting political feature of *The World Set Free* was a description of use of the new source of energy for explosives dropped from airplanes in a war with England and France against Germany and Austria. The "explosives" described by Wells were actually more like what we would envision as melt-downs of nuclear reactor cores as he did not anticipate the idea of a fast expanding chain reaction.

By 1915 decay schemes for various isotopes of radioactive elements were understood and the idea of quantitation was gaining wide acceptance especially after R.A. Millikan confirmed Einstein's photoelectric effect predictions in 1916. Nuclear processes should, of course, also be subject to quantization. The atomic model through 1920, envisioned an atom as made up of equal numbers of protons and electrons with half of the electrons residing in the nucleus as "nuclear electrons." Beta rays were nuclear electrons that had somehow been released and alpha particles presumably consisted of 4 protons and 2 nuclear electrons. Rutherford hypothesized a complex (three-dimensional) nuclear structure in which all the nuclear electrons were associated with protons on the periphery of the nucleus (as alpha particles).

Lisa Meitner

Lisa Meitner (1878-1968) received her doctorate as a student of Ludwig Boltzmann at the University of Vienna. Unlike Marie Curie, she was well connected and financed. Although there was a strong bias against women in physics in Eastern Europe, she managed to get a position with Max Planck in Berlin (1902). Otto Hahn (a chemist) accepted her as a colleague and together they discovered several elements and published papers in the years before WWI. In 1909, Lisa presented two papers on beta radiation. She had expected each element to produce a beta with a characteristic energy (as observed for alpha particles and x-rays), but the spectra had multiple lines and could not be immediately interpreted. Then in 1914, Chadwick, who was working with Hans Geiger in Berlin, reported that the beta spectrum of a mixture of ^{214}Pb and ^{214}Bi (at the time, this was called "radium B+C") was a *continuum*. Rutherford argued that all the beta-ray lines observed by Meitner, Hahn and their coworkers were from secondary electrons. Meitner would have to wait for an answer. So would Chadwick, as he happened to be in Germany when the war broke out in 1914 and was interned at the Ruhleben camp near Berlin during the war.

The Continuum Paradox

When the war came, Meitner briefly served as an x-ray technician, but by 1916 she was back in the laboratory and she and Hahn discovered more isotopes/elements. Returning to the issue of beta decay energies, Meitner was in the difficult position of trying to explain a continuous spectrum of primary beta-rays in a world where quantum effects were now widely accepted. Rutherford and coworkers were thinking that a nuclear gamma-ray might be causing the emission of the secondary electrons.

Meitner spent part of 1921 studying x-ray spectroscopy, which had been used to determine the energies of the inner orbitals for many elements by that time.[6] She hoped to use this knowledge to distinguish primary and secondary beta-electrons. She still expected a single, sharp primary beta line spectrum. Meitner developed a hypothesis that seemed very reasonable and for which she obtained conformation from several experiments. She reasoned that in beta-decay, the decay event occurred and the primary beta-electron with its full energy was produced. However, before leaving the nucleus, some of the betas produced gamma rays, which generated secondary betas from an internal photoelectric effect. Thus, the beta spectrum would include the full-energy primary beta and a series of secondary electrons that could be shown by their kinetic energies and the orbital energies from which they were ejected to be generated from the gamma (produced from the primary beta).

***

Another possibility for energy dissipation of an *electron* with very high kinetic energy (greater than 1022 keV) might be pair-production, to produce three particles (two electrons and one positron) with reduced kinetic energy. Collisions among these particles (including annihilation of one electron and the positron) might end up producing a spectrum of electrons with maximum kinetic energy just less than 1022 keV and a spectrum of low energy (blackbody) radio waves.

---

[6] Henry Gwyn Jeffreys Moseley 1887 – 1915.

***

Nothing is ever simple. About the same time that Meitner reached these conclusions, Charles D. Ellis (1895-1980), who had the luck of being interned with James Chadwick in German during the war and who then studied with Rutherford after the war, was studying some of the same isotopes by a very similar method and concluded the reverse: The gamma ray was generated first, impacted an electron in the lower energy orbitals and ejected it; then the electrons from higher energy orbitals fell into the lower energy levels and triggered beta decay. Ellis was drawn to this conclusion from indications that the gammas were being produced by primary nuclear events. Ellis went farther and suggested that the gamma emission actually triggered the beta decay; and (like Chadwick in 1914) Ellis was certain that the primary beta spectrum was a continuum (not distinct lines as expected from a quantum change).

Ellis appears to have come close to the truth with his analysis:

> Once the electron is separated from the neutron leaving behind a proton, *"The kinetic energy of the electron must be considered to depend on other factors"*... besides the condition that previously existed... *"the kinetic energy is possibly connected with the two facts [: (i)] that the nuclear field must vary considerably in distances comparable to the diameter of the electron and [(ii)] the electron*

> *cannot be considered as rigid[7] under these conditions."*

I would interpret Elise's comment as follows: imagine an electron suddenly appearing at an arbitrary point inside the nucleus of an atom (with some kinetic energy derived from a quantized event, and vector in an arbitrary direction), the potential energy of attraction that such electrons will experience in the nucleus will not be uniform, but rather it will depend upon the net positive charge between the electron and the center of charge of the nucleus (Gauss's law for a uniformly charged sphere). Moreover, for an electron (unlike the alpha particles and gamma photons), the particle-wave duality must be considered (i.e., the electron is not a rigid sphere). Of course, it was 1922, the concept of electrons as waves was in the air although Louis de Broglie (1892-1987) did not state it until 1924 and it was not demonstrated until 1927 (Davisson-Germer experiment). The primary beta electron would need to find an unfilled orbital and of course, there were none near the nucleus. Thus, regardless of its charge, it would be rejected from the region of the nucleus (Pauli exclusion) as a wave. Its trajectory and kinetic energy would undoubtedly be

---

[7] This seems to imply that the electron itself is "distorted" into some unstable (excited state), which accounts for some energy and presumably would be released upon collisions of the electron far from the atom.

impacted by the instantaneous distribution of charge in the nucleus and electron shells.

In the debate that followed, Ellis accepted the beta spectrum was a continuum and Meitner held out the view that it was actually somehow a misrepresented line spectrum. Meitner held to the analogies with alpha and gamma spectra and Ellis considered the electron to be different. Meitner's response was clearly stated:

> "*...in general it is found that by emitting radiation* [i.e., Ellis's beta-triggering gamma ray] *the electron falls to a more stable energy state.* [Ellis] *presumes that just the opposite occurs.*"

Here Meitner appears to take the "triggering" very literally (setting up a straw man that she could knock down). Clearly, an event like discharge of a gamma ray might eliminate a barrier (activation energy) that had been holding an electron in an unstable quantum state. She continued:

> "*In addition, the claim that the electron may leave the nucleus with any allowed speed, and not with a single speed defined by the energy state of the nucleus, is hardly an explanation for the existence of a continuous spectrum, it is only a description of it.*"

Meitner is correct that Ellis did not specifically state how his "two facts" affected the electron's kinetic energy. But she did not effectively rebut the issues that Ellis raised either. The disagreement between Ellis and Meitner became more personal: She criticized his spectra and his interpretation of the spectra (e.g., were they just poorly resolved line spectra?) and he treated her as an (irresponsible) assistant to Hahn.

It was soon easy for Meitner to show that gamma radiation was not always present when beta radiation occurred. She thought that she won that point after laboriously isolating isotopes that gave betas without gammas. But Ellis continued to argue that there was evidence for triggering gamma-rays (even though some may be as weak as x-rays).

When Meitner reported her results, the thrust of her paper was a rebuttal to Ellis. Ironically, she, thus, missed the novelty of what she had found. As Ellis had argued, secondary electrons are emitted by an atom when a high energy photon (x-ray or gamma-ray) collides with an inner shell electron and ejects it from the atom followed by the collapse electrons from higher orbitals into the lower energy orbital, which may be accompanied by release of photons (x-rays) representing the transitions from the outer orbitals to the inner orbitals. However, sometimes, the energy released by the collapse of the inner electrons is transferred directly to outer electrons, which are ejected from the atom as secondary electrons with kinetic energy. This *radiation-less* transfer of energy to electrons probably through the overlapping (interpenetrating) atomic orbitals was a novel observation by Meitner worthy of notice. In 1924, Pierre Auger noticed and focused on this phenomenon and the secondary electrons are now called Auger electrons (instead of "Meitner electrons").

In 1923, Meitner received unexpected support from the work of A. H. Compton, who showed that collisions of quantized photons with electrons in atoms resulted in a continuum of electron kinetic energies. Her view was that the primary beta (although Ellis might have argued a primary gamma) impacted the orbital electrons and produced secondary electrons with distinct kinetic energies while the primary beta(s) ended up with a continuum of energies. Regardless, by October 1925, Meitner determined that gamma rays were always emitted *after* nuclear decay. Specifically, *after* a change in nuclear state had occurred, release of the gamma ray was a way to remove energy from the *new* system. Ellis had reached the same conclusion and, in fact, proved it for beta decay. He wrote a very conciliatory letter to Meitner acknowledging that… "You were right!" Nonetheless, the issue of the continuum in the primary beta spectrum remained.

Ellis wrote:

> *"… We both agree that once the disintegration electrons are outside the parent atom, they are already inhomogeneous in velocity. We both agree that a quantized nucleus ought to give disintegration electrons of a definite speed, but*

> *whereas you think various subsidiary effects [e.g., Compton scattering] are sufficiently large to produce the observed inhomogeneity, I think they are much too small. ..."*

Political and economic conditions in Germany were in chaos in 1923-24, and it was 1926 before Meitner could gather the physical and mental resources to revisit the continuum problem. All the data to date was relative, absolute values of kinetic energy were not clear. She began a series of experiments in the Wilson cloud chamber to determine the range of beta-rays from ^{210}Pb (a.k.a., Radium D, RaD). Although she could identify, the two sets of secondary electrons from ^{210}Pb, no group of tracks attributable to mono-energetic primary beta electrons could be claimed.

Meanwhile, Ellis (who had left the concept of a gamma trigger far behind) studied ^{210}Bi (RaE), which has a pure beta decay with no gamma-rays, and confirmed that the primary beta electrons had a continuum of kinetic energies. Without gamma rays, there would be no Compton scattering as assumed by Meitner. Other experiments from the Cavendish Lab (K.G. Emeleus) yielded another important bit of information: If the continuum were caused by secondary electrons of any sort (e.g., Auger electrons caused by the primary beta), it would be expected that each disintegration would produce more than one beta electron; but the results indicate that each beta decay produced only one beta electron (1.1 +/-

0.1). This point had been assumed by both Meitner and Ellis and it was good that it was now tied down by Emeleus.

Meitner had her team confirm the Cavendish work and attempted to study the recoil of the nuclei. Because of the difference in the masses of the nucleus and the beta electron, equi-partition of the momentum means that the linear velocity of the nucleus would be tiny and almost all of the linear kinetic energy of the event ends up in the electron (i.e., there should be very little broadening of the beta peak). The nuclear recoil results were inconclusive.

***

One assumption made in all these studies is that the recoil acts along the center of mass of the electron and the nucleus. The possibility that the recoil of the electron might be opposed by a torque on the nucleus (nuclear spin) as opposed to linear nuclear translation appears to have not been considered. It is not clear how the angular moment of inertia of the nucleus would compare with the mass of the electron, but it might offer a partial solution to the question of a continuum of beta kinetic energy. Nuclear spin is also quantized, but there may be numerous closely spaced energy states for complex nuclei.

***

Ellis came up with a particularly interesting idea. He would use a calorimeter to determine the average energy of each beta decay. This was feasible only with ^{210}Bi (half-life 5.1 days) because it decayed rapidly and it did not have gamma or other radiation that would confuse the results. However, the polonium, which is produced from the bismuth, decays with a half-life of 139 days and produces alpha absorption. This liability was turned into an asset because it was used by Ellis and W.A. Wooster to accurately count the number of decays. It took several years to design and build the calorimeter and conduct the experiments. Ultimately, the calorimeter was designed to measure the *rate of heating* (not the absolute amount of heat generated). It was calibrated with radium and a thermocouple was used to measure the rapid increase in temperature (merely, 0.001 °C, which stabilized in about 3 minutes).

It is not entirely clear if Ellis and Wooster were planning (1925, *Proc. Camb. Phil. Soc.* 22:849-860) to measure *all* the energy released by each beta-disintegration (that would have seemed to be the point of the experiment), or just the energy carried by the electrons (which is what is observed in the beta-electron spectrum)[8]:

> "*This is to find the energy of the beta-rays from radium E. If the energy of every disintegration is the same, then the heating effect should be between 0.8 and 1.0 x 10^6 V per atom and the problem of the continuous spectrum becomes the problem of*

---

[8] Equi-partition of momentum, as discussed above, would mean that almost all the energy of the decay would be carried by the electron.

> *finding the missing energy. It is at least equally likely that the heating effect will be nearer 0.3 10 x 10^6 V per atom, that is, will be just the mean kinetic energy of the disintegration electrons."*

The first sentence talks about the "energy of the beta rays"; but the next sentence implies that the energy detected in calorimeter might equal that of the most energetic betas (presumably all the energy in the system), which implies that (as Meitner was arguing) some of the energy was being lost to other venues (e.g., the nucleus or perhaps chemical energy), which could be tracked down. It is, however, questionable whether or not the ultimate design of the calorimeter would allow energy sequestered in the nucleus or atoms of the sample to be measured. It appears that the calorimeter was designed and operated to measure *only* the energy contained in the ejected particles (not the energy retained in the nucleus/atoms): *Ordinary heat transfer from the nuclei (i.e., blackbody radiation from the solid sample to the lead absorber followed by conduction through the lead) to the thermocouple would be very slow whereas the ejected beta- electrons traversed to and through the absorbing material (1.2 mm of lead) instantaneously and were absorbed (Bragg peak) near the thermocouple (so that measurable temperature increases (0.001 ºC) could be detected within 3 minutes).* Because of the Bragg effect, it is possible that the temperature in the absorbing material was higher near the thermocouple than on the surface facing the sample. Overall, keep in mind that energy left with the nucleus (i.e., source), if any, does not appear to be accounted for; and, it is likely, this may have an impact on the interpretation of their results.

Their results were published in December 1927 (*Proc. R. Soc. London. A*, **117**:109-123). They found that the *energy of the ejected electrons absorbed by the source* was on average 350 +/- 40 eV as compared with 400 +/- 60 keV obtained from the spectrum obtained by Mr. Madgwick in the Cavendish lab. Thus, both systems seem to be accurately measuring the energy of the beta electrons (not necessarily of the entire system). They argued that if all the electrons released from the atom were at the highest energy electrons observed spectroscopically (1,050 keV); they would have observed a temperature rise consistent with 1,050 keV. Thus, the electrons released from the *sample* are indeed a continuum of energies. But, *without assuming that no heat was left in the bulk sample*, these data do not confirm that the primary process produces a continuum of energy. Nonetheless, Meitner and everyone else concerned about quantum theory and conservation of energy made the assumption that Ellis and Wooster had measured *all the energy* of every beta decay and were at a loss to explain what was going on. Meitner had her team repeated the experiment with minor modifications and came up with an average value of 337 +/- 20 keV for each beta decay (again assuming that the *beta electrons* represented *all the heat* produced in each decay). It was Meitner's turn to write a gracious letter, "… [Dr. Ellis] *you were absolutely correct in assuming that beta radiations are primarily inhomogeneous. But I do not understand this result at all.*"

***

And here the problem rested. Two of the best laboratories in the world (Cavendish and Kaiser Wilhelm) had beaten this problem for over 5 years and come to an agreement that current theories did not account for their results. One of the features that colored their thinking was that they gave little consideration of the nucleus as a physical object (although Ellis had offered that as a solution at one point). Most physicists of that day seemed to treat the nucleus as a point mass and point charge (rather like the electron). Moreover, because the energy of nuclear events is so much larger than chemical events, no one showed any interest in considering the chemical energy associated with turning a bismuth atom into a polonium atom in a chemical matrix. Finally, at this point, the neutron had not been discovered. This situation made the entire issue of beta-rays from the nucleus a questionable proposition. Nonetheless, before considering scraping such a fundamental concept as conservation of energy, it would seem that much more work would be needed.

> I have produced an explanation for these observations that do not require any exotic particles or explanations. I will save the details for later but the problem really boils down to the fact that everyone was treating the nucleus as a point mass/point charge. Here let us continue with the historical record.

Chapter 5. Pauli's Particle

Rutherford, Chadwick, Ellis, Meitner, Hahn and a host of others had tried valiantly to explain beta decays and the continuum of kinetic energy observed in the ejected electrons. It was obvious that the electrons from the nucleus must also have something to do with the "nuclear electrons" and "neutral doublet" ideas that were circulating along with George Gamow's (1904-1968) liquid drop model of the nucleus (circa 1928). Gamow had recently explained how alpha particles could escape the nucleus by quantum mechanical tunneling (1928. *Zeit. f. Phys.* 51:204-212). Gamow's calculations caused him to doubt the existence of "nuclear electrons" by the mid-1930s.

Pauli in Turmoil

Meanwhile, Wolfgang Pauli (now 30 years old) had spent the last 8 years lecturing at the University of Hamburg and had not taken direct part in any of the debates concerning beta electrons, nuclear structure, etc. that important physicists had been involved with for the last 10 years. The only thing, he had developed during this time was a reputation for breaking laboratory equipment... dubbed the "Pauli Effect." He had also used the undisciplined life of a college professor to indulge his penchant for late night relaxation in bars and cafes. Like many professionals with Jewish heritage, he felt pressure in Germany in the late 1920s. Einstein was a lightning rod, but he was too prominent for the Nazis to intimidate at the time. But Pauli, whose greatest claim to fame was glorifying Einstein, was an important and easier target for the Nazis. He, thus, found a position as a professor of theoretical physics at the Federal Institute of Technology in Zurich, Switzerland in 1929. This was not a move up; it was an escape.

Overall, Pauli's life was in turmoil. In 1927 his mother committed suicide. His father quickly remarried a woman who Pauli detested (perhaps his new step-mother was the cause of his mother's suicide). Upon leaving Berlin, Pauli apparently impulsively married a woman named Käthe Margarethe Deppner, who is uncritically described in history as a "dancer," on 23 December 1929. Most likely Ms. Deppner worked the Berlin cabarets of the late 1920s. To add details to her situation, many young women had been left destitute by the hyperinflation of the 1920s and had come to Berlin where people with solid incomes lived and played. These girls ended up working in clubs that satisfied every sexual fantasy for a price. Ms. Deppner left Berlin with Pauli, but he was probably not the fun guy with money that she expected. She had no appreciation for his profession and he drank heavily. She divorced him on 29 November 1930 in Vienna.

Pauli was desperate to revive his career. He was aware that a meeting of the most important physicists in German was taking place in Tubingen, not that far from Switzerland; but he was not invited. So, in a depressed state he abandoned all pride and addressed a letter to the combined gathering, 5 days after his divorce (4 December 1930). The letter was delivered to Lisa Meitner and Hans Geiger apparently by telegram from Zurich:

> "Dear radioactive Ladies and Gentlemen!
>
> As the bearer of these lines, for whom I ask your gracious attention, I have, faced with the "false" statistics of the N-14 and Li-6 nuclei as well as the continuous beta-spectrum, stumbled upon a desperate remedy. Namely, the possibility that in the nucleus there could exist electrically neutral particles, which I will call neutrons, which have a spin of one-half and obey the exclusion principle and in addition also differ from light quanta in that they do not travel at the speed of light. The mass of the neutron must be of the same order of magnitude of the electron and in any case not larger than 0.01 proton mass. The continuous beta

spectrum would then be understandable assuming that in beta-decay a neutron is emitted along with the electron in such a way that the sum of the energies of the neutron and electron are constant...

At the moment, I do not trust myself enough to publish anything about this and turn to you, dear radioactives, of how one might experimentally prove such a neutron if its penetration power is similar to or about ten times that of gamma radiation.   I admit that my remedy may at first seem only slightly probable, because if neutrons do exist, they should certainly have been observed long ago! But, nothing ventured, nothing gained, and the gravity of the situation with the continuous beta spectrum is illustrated by a statement of my respected predecessor in this office, Herr Debye, who told me recently in Brussels: "Oh, it is best not to think about it at all, like the new taxes!" Thus, one should seriously discuss every means of salvation. Therefore, dear radioactives, test and decide!

Unfortunately, I cannot appear Tubingen in person, since I am indispensable here due to a ball which will take place the night of December 6 to 7 in Zurich.

With many greetings to you all,
your most humble and obedient servant,
W. Pauli

The Prefect Theory

As Meitner had told Ellis, this is not so much a solution of the problem it is description of the problem. It is not clear why anyone took it seriously except that they had apparently run out of ideas themselves both in the realm of figuring how to account for the mass in the nucleus and the continuous beta spectrum. Pauli was suggesting that the nuclear electron somehow recoiled off of a tiny mass of neutral material, which had not been detected because it does not interact with matter (at least within the experimental range of that time). And, no one asked how the massless particle recoiled off the electron if it does not interact with (any other) matter. This recoil (like Compton scattering) would provide a continuum of electron energies…. Meitner had already considered that idea, but gave up when beta-decay was observed without gamma-rays or secondary electrons. Since no one had yet proven the existence of Rutherford's "neutral doublet" which seemed to occupy the nucleus, it was hard to object to Pauli on the grounds of the mass of the neutron. So, the impossible idea hung around.

Pauli had the perfect theory: it could not be proven wrong experimentally.

Pauli had gained attention without doing any experiments or offering any testable solutions. Indeed, his test was a classic example of "negative evidence." You can never prove the proposition that "the particle does exist but you cannot detect it" to be wrong. You can only attempt to do so, and by random error you may think you proved it does exist. Pauli may have realized that he could not be proven wrong when he challenged the experimentalist to "...test and decide!" If he were lucky, he would eventually be validated.

But Pauli's idea really did not take root among the trained physicists of the day in Europe. Nonetheless, he toured the US and presented the idea at a news conference in Pasadena, California (16 June 1931). *The New York Times* (17 June) accepted his absurd claim and made it seem as though it was a main-stream idea:

> "A new inhabitant of the heart of the atom was introduced to the world today when Dr. W Pauli of the Institute of Technology in Zurich, Switzerland postulated the existence of particles or entities which he christened 'neutrons'."

Fermi's Biggest Mistake (1934)

Chadwick, following the work of others, identified what we today call the "neutron" in 1932. It was obvious from the mass of the (real) neutron that no mass was left over for Pauli's particle. Thus, when Pauli finally went into scientific publication, he stipulated that his particle was assigned zero mass (or at least no mass that could be detected). This proclamation made the supposed particle even that much more difficult to find.

There was still no recognized theory for the nature of beta decay. Finally, in 1934, Enrico Fermi (1901-1954) developed and published a theory of beta decay assuming the reality of a mass-less charge-less particle, which was demoted to the title 'neutrino' (little neutron). Fermi began by agreeing that only particles with large mass (i.e., short de Broglie wave lengths) could exist in a small space like the nucleus (10^{-15} m). He then describes the inter-conversion of neutrons and protons as a change in quantum state (rho) that must be accompanied with creation or annihilation of electrons and neutrinos. A chemist would say that the conversion of neutrons into protons and electrons is a reversible equilibrium process. Fermi introduces several pages of linear algebra where (after some simplifying assumptions) he develops energy operators. After arriving at the conclusion that the energy of the system is the energy of the heavy particles and the light particles, he begins the application to beta decay. In section V of his paper he states,

> *"A beta decay is the process by which a nuclear neutron changes into a proton at the same time an electron, which is observed as a beta ray, and a neutrino are emitted by the described mechanism."*

But he actually has not described a mechanism. He goes on to define one state in which the neutron exists and another state in which the proton, electron and neutrino exist. In part VI, Fermi concludes that for beta decay to occur, the energy of the nucleus with the neutron must be higher than the energy of the nucleus with the proton. In section VII, Fermi concludes that for a rest mass of the neutrino equal to zero, the high kinetic energy end of the beta spectrum will be concave upward (his Figure 1) and notes

> "...the greatest similarity to the empirical curves is given by the theoretical curve for mu (mass of the neutrino) equal to zero."

It might be mentioned that if the neutrino did not exist, you would have the same effect. In parts VIII-X, Fermi derived the rest of the shape of the energy spectrum, which is consistent with the experimental shape reported by Ellis and Wooster (1927 based on data from Madgwick).

But note that Ellis and Wooster (1927) made a point of the fact that the Cavendish lab equipment artificially eliminated low-kinetic energy betas because ...

> "...owing to the necessity of covering the opening of the ionization with mica, slow electrons, if present, will not have been measured."

They also corrected each data point by multiplying by the factor capital-beta-squared, ... "...where capital-beta is the velocity [of the electrons] in terms of the velocity of light [i.e., v/c]." These two factors should surely drive down the intensity of the "corrected" experimental beta-ray beam as the kinetic energy approaches zero. In modern data sets (for example Florian Marhauser TU

Darmstadt, Relativistic Heavy Ion Physics Seminar 10.11.2005, TU Darmtadt, p. 3), the data points are shown and dip only slightly below 200 keV (i.e., about 85% of max intensity at 10 keV kinetic energy). These results are consistent with the model provided by Fermi. But the model provided by Fermi says very little about the neutrino except that its mass is indistinguishable from zero.

Overall, Fermi's paper (e.g., translated by Fred L. Wilson, 1968. *Amer. J. Phy.* 36(12):1150-1160) should not be taken as a rationalization of Pauli's proposal. Indeed, his model only corresponds with experiment if the mass of the neutrino is zero (or the neutrino is non-existent).

Since, neither the electron nor the neutrino is presumed to exist in the initial state of the nucleus, the overarching question is how can the mass-less, charge-less neutrino (which must be some sort of a photon, though perhaps not an electromagnetic photon) carry away momentum and (kinetic) energy. In modern texts, we find the "W-boson" which is produced with the collapse of the neutron and then produces the electron and the neutrino (actually an "electron antineutrino" in modern-speak). This system avoids the difficulty of explaining how all the recoil energy absent from the electron ends up in the neutrino (and not e.g., in the proton). However, it is not clear, how the W-boson separates from the proton without producing some recoil effect or charge separation effect.

Chapter 6. The Real Neutron 1932

Irène Joliot-Curie

Although Pauli had somewhat muddied the water, with his calling his particle the 'neutron', few scientists confused it with the neutron long pursued by Rutherford and Chadwick. The honors for unlocking this problem actually go to none other than the daughter of Marie and Pierre Curie. Irène Joliot-Curie (1897-1956) had inherited or absorbed her mother's drive and scientific interest. During WWI, they had worked together in mobile x-ray laboratories trying to save the lives of allied soldiers...and they would both pay the price for the high doses of x-rays they received. Nonetheless, Irene married (her husband took her surname) and launched into a study of nuclei in the late 1920s.

In 1932, they followed up the 1930 experiments of German physicists (Walter Bothe and his student Herbert Becker) who had bombarded light elements (Li, Be and B) with alpha particles and observed highly penetrating radiation (not affected by electric fields) that they though was high-energy gamma radiation. Joliot-Curie repeated the German experiment with a strong alpha source and discovered to their surprise that the radiation coming from beryllium (when bombarded with alphas) ejected protons from hydrocarbon wax. They mistakenly explained this as sort of a Compton scattering effect of high energy photons.

James Chadwick

But their observations quickly made it to James Chadwick in England. In conference with Rutherford, Chadwick dismissed the Joliot-Curie idea and correctly interpreted the experiment as demonstrating the neutron, a neutral particle with a mass a little greater than the proton, just as Rutherford had predicted in 1920.

The Atom

With the discovery of the electron, proton and neutron, the Bohr-Rutherford model (as promptly clarified by Heisenberg and Schrodinger) provided a modern understanding of the atom, at least for physicists. The chemists were not fully satisfied until Linus Pauling came up with the idea of hybrid atomic orbitals, which explained the geometry of molecules. But since chemistry is all about the electrons, chemists have had no need for more details about the atom's nucleus than to acknowledge that there are isotopes of each element, which account for the irrational atomic masses and show up in rotational and vibrational spectra and mass fragmentation patterns.

For the most part, the physicists have been content to try to subdivide the elementary particles into smaller pieces. Regardless, I do not find that the neutrino plays any fundamental role in explaining atomic structure or nuclear reactions. But a lot of money has been spent (is being spend) studying neutrinos and a lot of physicists seem to have staked their reputations and careers on understanding this particle created by an arrogant guy probably in a drunken state.

I can contain myself no longer. Let's end this nonsense right now.

May I Interject!!

As stated in the summary, everyone into the 1930s seemed to assume the nucleus was a point mass and point charge. As a point mass, it has no angular momentum (radius = 0) and no internal volume. Any line of force applied to it goes through the center of mass and the full nuclear charge is experienced by any particle in the vicinity. There is a simple solution to the beta-electron kinetic energy continuum. And here it is. Gauss's law requires that the net charge (field) experienced by a particle inside a sphere of homogenous charge is only that fraction of the charge within the radius of the particle.[9]

Thus, depending upon where *in the nucleus* the electron is launched from, it will experience a different nuclear charge. Chemists are very familiar with the phenomenon of electron "shielding" outside the nucleus. Basically, the nuclear charge experienced by an electron outside the nucleus is given by

$$Z_{eff} = Z - \sigma$$

Where σ is the shielding constant, which is the electron density inside the average orbital radius of the electron and Z is the full nuclear charge (since the probability of finding a proton outside the Bohr radius is nil).

---

[9] Just like gravity. If you made it to the center of the earth, you would be weightless.

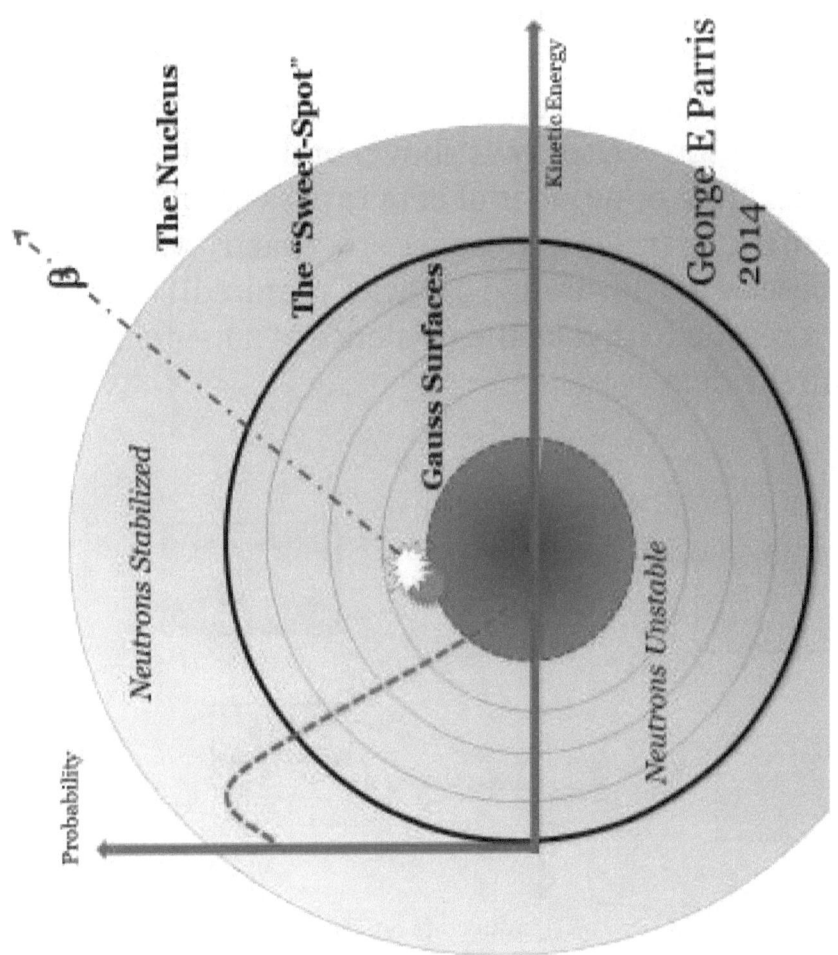

Since electrons may be launched anywhere inside the nucleus by the spontaneous transition of a neutron to a proton and electron, and *this is a quantum jump*, the kinetic energy carried away by the electron is reduced by the potential energy associated with the effective nuclear charge. The probability of a neutron undergoing this transition at radius "r" inside the nucleus is proportional to the volume (charge) enclosed within the radius (r) compared to the entire radius of the nucleus. The result of this situation is exactly as shown in the figure: The theoretical plot of number of beta rays as a function of their kinetic energies in this model is exactly what has been observed. No strange particles required!! There was never a reason to invoke the existence of a neutrino to explain the data!

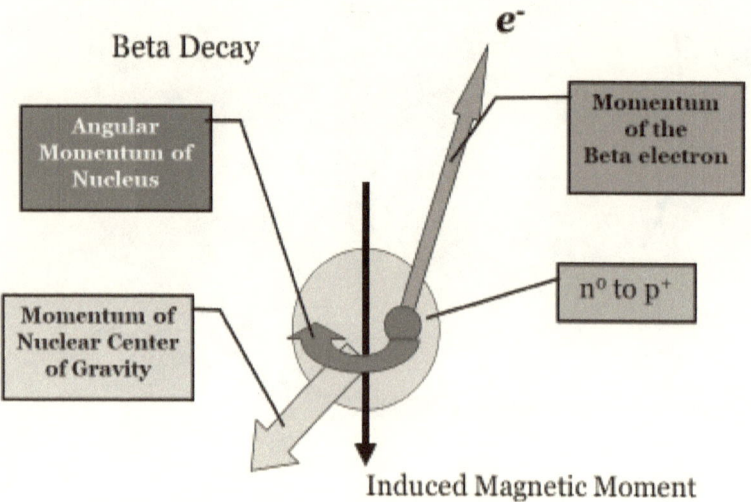

There is also another issue. The momentum applied to the nucleus by the release of the electron will rarely go directly through the center of the nucleus.

I am going to guess that there is some nuclear angular momentum (which is quantized) and must be satisfied before the neutron can disintegrate. These changes in "spin" may actually be observable by changes in nuclear magnetic moments. [10]

There are some implications of this model to the structure of the nucleus. In particular, stable isotopes would presumably have a majority of the neutrons near the surface of the nucleus where they knit the surface protons together in a mosaic of *nascent alpha particles*. The protons on/near the surface would, of course, experience more repulsion (higher Z_{eff}) than protons nearer the center of the nucleus and the neutrons are there to glue them together. I believe this is essentially the structure that Rutherford had in mind (see page 34).

There must be a minimum number of neutrons needed to hold the nucleus together; but if the number of neutrons becomes excessive, the nucleus will be prone to disintegration by beta decay from the inside or alpha decay from the surface (alphas being the preferred positively charged particle ejected because of its inherent stability and simplicity).

Qualitatively, these arguments seem reasonable. But I convinced myself that I was really on to something when I did some calculations:

---

[10] I would like to know what the nuclear magnetic resonance of pure tritium oxide (T_2O) looks like? Actually ^3He derived from ^3H.

Calculation of E_{max} for ^{210}Bi

While neutrons can be indefinitely stable in most nuclei, it turns out that free neutrons spontaneously decay with a half-life of slightly over 10 minutes (611.0 ± 1.0 s).[11] This fact may have contributed to the difficulty in discovering them. It also suggests how the neutron can disintegrate into a proton and an electron (beta particle, i.e., electron with kinetic energy):

$$n^0 \rightarrow p^+ + \text{free } e^- + \text{kinetic energy (782.343 keV)}$$

The <u>kinetic energy released</u> is based on lost mass and Einstein's equation.[12]

If we make the assumption that the neutron (in a nucleus) exists until the electron is separated from the proton and

---

[11] This gives a mean-lifetime of 881.5 ± 1.5 s.

[12] The beta decay of 210Bi is what is always cited. Has anyone observed the spectrum of kinetic energies of electrons from free neutrons (e.g., in cloud chambers)? I assume that (as Meitner assumed) the primary event is quantized. Presumably, the electron would carry away most (but not all of the kinetic energy) and the electric potential of separating the charges would reduce the actual observed kinetic energy of the electron.

enters the first Bohr atomic orbital of hydrogen[13] (at which point the electron and proton have achieved separate identities and Coulomb's law begins to apply), the process can be broken into two steps:

(1) Neutron inflation to Hydrogen atom (loss of mass with <u>energy release</u>)[14]

(2) Hydrogen atom to free Proton and Electron (Coulombic <u>energy uptake, endothermal</u>)

$$n^o \leftrightarrows {}^1H_1 \rightarrow p^+ + e^-$$

The energy needed for the second step is just the ionization potential of the hydrogen atom (1312 kJ/mole or 13.64 keV/atom). (Yes, the Bohr radius of the hydrogen atom is much larger than the nucleus of any element, but the electron density is still centered on the proton, so the net negative charge is centered at the position of the proton.) Since the total energy for the process (step 1 plus step 2) is 782.343 keV we can calculate the energy for step 1 (which should be independent of the nuclear isotope) as follows:

---

[13] This is equivalent to saying that the electron tunnels out of the nucleus.

[14] I suggest calling this "neutron inflation."

Overall:

$$-782.3 \text{ eV} = \text{step 1} + 13.6 \text{ eV}$$

(negative sign indicates exothermal)

$$\text{step 1} = -782.3 \text{ keV} - 13.6 \text{ keV} = -795.9 \text{ keV}$$
(exothermal)

Now, if an element like bismuth (Z = 83) had the process happen at the center of its nucleus (based on the Gauss's law) we would expect a beta-ray with energy equal to 795.9 keV *plus* whatever additional nuclear binding energy was achieved from going from

$$^{210}\text{Bi} \rightarrow {}^{210}\text{Po}^+ + \beta^-$$

Thus

$^{210}\text{Po}_{84}$	1645228 keV
$^{210}\text{Bi}_{83}$	1644849 keV
	379 keV (exothermal)[15]

Thus, we would expect a KE_{max} for a beta from the center of the nucleus of ^{210}Bi to have kinetic energy:

---

[15] Using data found at
http://onlinelibrary.wiley.com/doi/10.1002/9783527618798.app2/pdf

$$KE_{max}(^{210}Bi) = 796 \text{ keV} + 379 \text{ keV} = 1175 \text{ keV}$$

Exothermal sum of energy from decay of the neutron and change in the nuclear binding energy. Then subtract the energy needed to ionize the incipient hydrogen atom to calculate the observable maximum KE of the beta ray

Calculated Observable KE_{max} (^{210}Bi) = 1175 − 14 = 1161 keV

Remember this is the relatively rare neutron decay that happens only at the nucleus of the atom (^{210}Bi). The actual experimental maximum KE is 1.16 MeV.

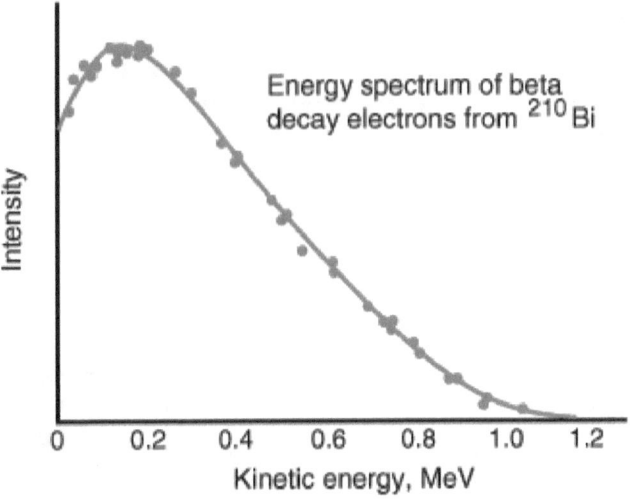

https://warwick.ac.uk/fac/sci/physics/research/epp/exp/detrd/amber/betaspectrum/ Experimental energy spectrum for decay electrons from ^{210}Bi, From *G. J. Neary, Proc. Phys. Soc. (London),* **A175**, *71 (1940).*

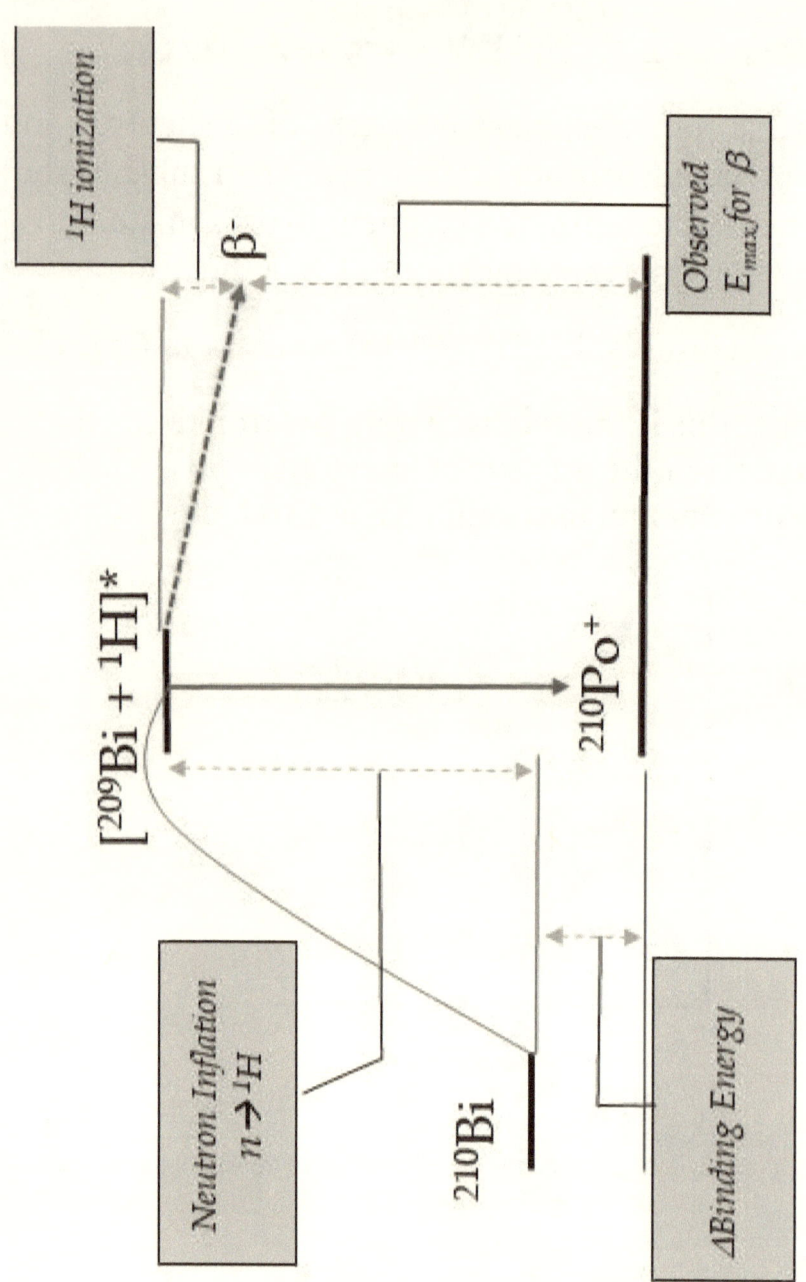

In contrast, if the neutron inflated to a proton and an electron *at the periphery of the nucleus of polonium* (Po_{84}) (assuming nuclear radius of 15×10^{-15} m or 1.5×10^{-4} A) yielding 1156.7 keV of energy, there would not be sufficient energy for the beta-electron to escape from the nucleus based on a calculation of the columbic potential:

Affinity potential = $[1.39 \times 10^3$ kJ-A/mole$] [(1)(84)/1.5 \times 10^{-4}$ A$]$

$$= 7.784 \times 10^8 \text{ kJ/mole}$$

$$= 8.064 \times 10^6 \text{ eV} = 8064 \text{ keV}$$

That is, 8064 keV (the potential energy holding the electron in the nucleus) is greater than 1175 keV (the energy generated by inflation and binding energy) so the beta would not escape from the nucleus (i.e., it would be forced to return to a proton reforming a neutron).

If we work this backwards to calculate the *maximum radius (i.e., distance from the center of the nucleus) at which the beta electron will escape from the nucleus*, we need to estimate the charge inside that radius:

1175 keV =

(1.39 × 10³ kJ-A/mole) X

(1.04 × 10⁻² eV/kJ/mole) X

[(1)(Gauss charge)/r]

Assuming that the positive charge is homogeneous in the nucleus,

Charge proportional to volume = {constant} $(4/3) \pi r^3$

Where the constant is approximately the nuclear charge of the heaviest elements:

Z of heavy element ~ 100 =

{constant} $(4/3)(3.14)(1.5 \times 10^{-4} \text{ A})^3$

{constant} = $[23.89/3.375 \times 10^{-12}] \text{A}^{-3}$ = $7.077 \times 10^{12} \text{ A}^{-3}$

Thus,

Gauss charge = $\{7.077 \times 10^{12} \text{ A}^{-3}\}(4/3)(3.14) r^3$

$= (2.963 \times 10^{13} \text{ A}^{-3}) r^3$

Thus, solving for the maximum Gaussian radius (r_g) at which a beta electron can escape from the nucleus (of total radius about 15×10^{-15} m):

1175 keV[16] =

$(1.446 \times 10^{-1} \text{ eV-A})[(1)((2.963 \times 10^{13} \text{ A}^{-3}) r^3)/r]$

---

[16] These calculations are of course dependent on the change in binding energy associated with the change in the nucleus (see above). Thus, these calculations are only relevant for ^{210}Bi.

$1175 \text{ keV} = [4.283 \times 10^{14} \text{ eV A}^{-2}] \, r^2$

$r_g = ([1175 \times 10^3 / 4.283 \times 10^{14}] \text{ A}^2)^{1/2}$

$r_g = 5.24 \times 10^{-5} \text{ A} = 5.24 \times 10^{-15}$ m

Thus, *in the case of* ^{210}Bi (nuclear radius ~15 × 10⁻¹⁵ m), the Gaussian radius for beta escape is about 1/3rd of the radius of the nucleus. Thus, we can conclude, that the beta electron can only escape if it is generated near the center of a nucleus (where the net Gauss charge is low). This result suggests a basis for the stability of the bound neutrons in most nuclei compared to a free neutron.

In this hypothesis, the neutrons of the nucleus reside either in the "sweet spot" (where they can decay) or in the remainder of the nucleus (where they are stabilized by the inability of the electron to escape). The probability of beta decay is related to the relative size of the "sweet spot" in the center of the nucleus where the Gaussian nuclear charge is low enough to allow the neutron to escape as a fraction of the entire nuclear volume.

If the "sweet spot" is divided into a series of thin shells of thickness dr (e.g., on the order of the radius of a neutron, 0.8×10⁻¹⁵ m), the volume of each shell will increase as "r" is increased. Thus, the shapes of the beta-electron energy spectrum (i.e., intensity as a function of kinetic energy) should follow a function that looks like

$$V(r-dr) = (4/3)\pi[r^3 - (r-dr)^3]$$

very much like the observed beta spectra (i.e., most of the electrons are emitted with low energy and the number of electrons declines to zero as the kinetic energy approaches its maximum (1175 keV).

Of course, near the periphery of the "sweet spot," the probability of beta decay falls off as the neutron (with finite size) is partially outside and partially inside. The ratio of the declining part of the curve to the ascending part of the curve should be related to the relative size of the "sweet spot" and the neutron. In the case of ^{210}Bi, that would be 0.8/5.24 = 0.15 based on calculations here. Based on the experimental curve (above) I calculate 0.15.

Calculation of Half-life of ^{210}Bi Beta Decay

In the case of ^{210}Bi, the fraction of active volume is estimated (see above) as $(5.24/15)^3$ = 0.043. For ^{210}Bi (atomic number 83) there must be 127 neutrons in the nucleus and *if they are randomly distributed* (as in the liquid drop model) we would expect 0.043 x 127 = 5.4 neutrons in the "sweet spot" at all times. Given that the half-life of a free neutron is 611.0 ± 1.0 s, we might predict that the half-life of ^{210}Bi would be 611/5.4 = 112 s. The half-life is actually about 5.0 days, which would imply a much smaller ratio "sweet spot"/nuclear volume.[17] The corresponding rate constant is 6.14 x 10^{-3}/s, which might be taken as the Arrhenius frequency factor.

This calculation (above) is based on uniform distribution of neutrons in the nucleus. If as supposed in the shell model of the nucleus, free neutrons are preferentially concentrated on the surface of the nucleus, that would lengthen the half-life. Either a free neutron would have to penetrate into the structured core of the nucleus or a paired neutron in the core would have to separate from its partner (interpretable as an activation energy).

Obviously, this is a chemist's interpretation of nuclear kinetics. The enormous energies involved in nuclear binding can explain the absence of the effect of normal thermal temperatures on nuclear decay. But, still there should be some explanation why a decay process that has a half-life of only a few minutes, is not affected by normal thermal temperatures. One explanation might be that while on the atomic scale these reactions are strictly first-order, in the nucleus perhaps they are higher order with a very low Arrhenius frequency factor (i.e., high entropy of activation). For example, suppose the nucleus in primarily composed of alpha particles, in the case of ^{210}Bi that would be 83/2 = 41 alpha particles accounting for 82

---

[17] I believe the Gamow-Teller half-life calculated for ^{210}Bi$_{83}$ is "greater than 100 seconds".

of the 127 neutrons, leaving 45 unassociated neutrons and one extra proton.

Then suppose there is a second order reaction is required between the proton and a free neutron and it must be in the sweet spot to be productive. The probability of this from random distribution alone becomes:

$(41 \times 0.041)(1 \times 0.041) = 0.0689$

$611 \text{ s}/0.0689 = 8865 \text{ s}$

Now, suppose I needed three neutrons and a proton in the sweet spot to form a *nascent alpha* particle:

$$n_3 p^+ \rightarrow \text{alpha}^{2+} + e^-$$

The probability goes to something like (I'm not a statistician):

$41 \times (0.041)^3 \times (1 \times 0.041) = 0.00012$

And the half-life goes to $611/0.00012 = 5.57 \times 10^6 \text{ s} \approx 60$ days

This is, of course, speculative manipulation of the numbers, which I would submit is not as absurd as invoking a mass-less/charge-less particle that carries away kinetic energy.

In summary, the paragraphs above explain (i) the continuity of the energy spectrum of beta decay, (ii) the shape of the beta decay spectrum, (iii) the maximum energy of beta decay and (iv) provide rationales for calculating the half-life for beta decay, *without mentioning the neutrino or invoking any particle with strange properties.*

The Origin of Neutrons

Neutrons appear to be a metastable quantum state in which the electron is in the n=0 (principle quantum number zero quantum state.

$$\Psi(0,0,0)$$

This state is created when hydrogen is placed under immense pressure in massive stars, literally hydrogen atoms are crushed ultimately to neutrons rather than metallic hydrogen. As described above, the neutrons are then captured and stabilized in the nuclei of elements (produced during supernova) by the electrostatic and nuclear binding energies.

It would be interesting to see if the frequency of inflation:

$$\Psi(0,0,0) \rightarrow \Psi(1,0,0)$$

is thermally sensitive. If it is not thermally sensitive, then it must have some statistical (quantomechanical) basis.

∙∙

***

In case you were asleep (or simply did not read between the dotted lines), the paragraphs above explain (i) the maximum energy of beta decay, (ii) the continuity of the energy spectrum of beta decay, (iii) the shape of the beta decay spectrum and (iv) the half-life for beta decay, **without mentioning the neutrino or invoking any particle with strange properties.**

***

Fermi's Second most Embarrassing Error

The story of the neutron would not be complete without mentioning that in the 1930s almost everyone started shooting neutrons at various elements in the expectation of making higher mass elements. Enrico Fermi was the star in this field claiming (1934) to produce various trans-uranic elements; in fact, he was awarded the Nobel Prize in 1938 for this discovery ("his demonstrations of the existence of new radioactive elements produced by neutron irradiation"). The literature reflected this bonanza of new elements and isotopes until it was realized that the neutron was actually catalyzing nuclear fission.

The chemist Ida Noddack (1896-1978) had criticized Fermi's work in 1934: rather than make larger nuclei

> "...it is conceivable that the nucleus breaks up into several large fragments, which would of course be isotopes of known elements but would not be neighbors of the irradiated element"

But she never received the attention she deserved for suggesting that fission rather than fusion was happening.

Thus, the neutron was the catalysts that H.G. Wells had speculated in 1914 would open the door to rapid release of nuclear energy for helpful and potentially devastating purposes. It was the careful work of chemist Otto Hahn, who identified Ida Nodack's "isotopes of known elements," that ended the trans-uranic party and left Fermi and the Nobel Prize committee with silly grins on their faces.

Once the fission cat was out of the bag, it only took a few days at the dawn of WWII for Niels Bohr to realize that a chain reaction of nuclear fissions and nuclear bombs would be possible (29 April 1939).

Chapter 6. Confirmations of Pauli's Particle

It is Impossible to Prove a Negative

Pauli dared the experimentalists to try to find his massless, charge-less highly penetrating particle. It carries energy and spin without mass or waves. What we will find is a series of complex experiments (based on assumed physics of the neutrino) that depend on statistical analysis of very small numbers of events. These experiments are generally not readily replicable by others because they require unique and expensive and time-intensive facilities and procedures. The results are generally right at the limit of reliability. The really bizarre problem is that the neutron detection events that happen inadvertently frequently lead to unexpected results: (i) the sun is not working right, (ii) neutrinos travel as fast as the speed of light, (iii) neutrinos have mass, (iv) neutrinos travel faster than the speed of light, (v) neutrinos have mass that fluctuates, (vi) neutrons have other strange properties. People publish this stuff with a straight face.

I am reminded of a quarrel that developed in 1923-27 between Rutherford and Chadwick (the Cavendish Lab) and Hans Pettersson and Gerhard Kirsch (the Radium Institute in Vienna, Austria). The dispute boiled down to the ability to distinguish alpha particles and protons by the intensity of their scintillations on zinc sulfide screens. The differences were in the eye of the beholder. The award of the Nobel Prize to Fermi in 1938 for his creations of transuranic elements, when in fact he was observing nuclear fission also comes to mind. Nonetheless, let's go into the tangled history of experimental detection of neutrinos a little bit.

Cowan and Reines

Clyde Cowan (Jr, 1919-1974) and Frederick Reines (1918-1998) met in 1949 at Los Alamos. They realized that nuclear reactors should be large sources of electron antineutrinos based on the beta decay that was present and from this they devised an indirect method of observing the neutrino. The idea was to have the electron antineutrino react with a proton (i.e., water was a convenient source of protons) to produce a neutron and a positron (i.e., antimatter electron, same mas as an electron but a positive charge). The positron and an electron (relatively well understood particles) would immediately collapse into a gamma ray (the reverse of pair production) and the slow neutron would soon thereafter be captured by a nucleus (a well-known process from the work with neutrons in the 1930s) to produce a heavier isotope and also yield a gamma ray. The coincidence of these two gamma rays in time would signal the presence of an electron antineutrino. It was a brilliant idea. They set up their tank of water with gamma ray detectors (i.e., scintillators in the water with photomultiplier tubes on the outside) and timers near a powerful reactor at Los Alamos calculated to produce a neutrino flux of 5×10^{13} neutrinos per second per square centimeter. The experiment did not produce interpretable results.

They added cadmium salts (cadmium is a strong neutron absorber) to the solution and moved the experiment first to US Department of Energy reactors at Hanford, (Washington) and finally Savanna River (South Carolina).

Cosmic rays were deemed to be the biggest problem creating background, against which the detection of the timed flashes of the required gamma rays were not detectable. So, the set up at Savanna River was moved 12 meters underground. Over months of observation, they concluded that they detected about 3 neutrinos per hour in their detector. Shutting down the reactor reduced the number of detected events. (*Science*,1956:124(3212):103-104).

You will notice that the experiment ended and was published when positive results were obtained, not when negative results were obtained.

The Neutrino becomes and Embarrassing Particle

But, the definition of a neutrino seems to be whatever it takes to make the experiment work. Recall that Pauli and Fermi claimed the neutrino should have no mass and travel slower than the speed of light.

George Parris Copyright Claimed August 2019

***

Apparently, they were wrong:

> The neutrinos first attacked science by shutting down the sun. Well not literally, but that is what people feared between 1960 and 2002. It turned out that a quick calculation revealed that the sun should be a more powerful source of neutrinos on earth than the reactors that Cowan and Reines used to demonstrate their existence.[18] Based on observations, we were receiving only a fraction of the neutrino flux that the solar physics suggested we should be experiencing. Oops, looks like the sun has stopped working...that could be bad for the planet. But, the sun kept ticking and by 2002, the "solar neutron problem" was resolved by experiments at the Sudbury Neutrino Observatory (Canada), which used Cherenkov radiation from high energy electrons produced by neutrino interactions in 1000 tons of heavy water in a 2-meter diameter plastic tube in the bottom of a deep mine shaft, to show that solar neutrinos oscillate among three types. Apparently, there are three different types of neutrinos (electron, mu and tau) and a neutrino created in one state will *oscillate* among all three different states because of its mass (yes, mass) as it propagates thorough space. In the

---

[18] What would this do to their original conclusions?

original experiment (the one that caused the solar problem) only one type of neutrino was being monitored. So, there must have been more neutrinos, we just did not see them.

By the way, neutron oscillation (at least mu electron oscillation) was confirmed in the summer of 2011 by observing 6 "electron neutrino-like events" at T2K when background was only 1.5 +/- 1.8. In 2013, this event was reaffirmed.

***

Supernova1987a was quite an event 168,000 light years from earth. It is calculated to have released about 10^{58} neutrinos and presumably earth got its share. There were four neutrino traps working on earth at that time. First the Mount Blanc liquid scintillator registered a burst of 5 neutrinos. Then, three hours later three other neutrino traps registered a burst of neutrinos over a period of time estimated to be 13 seconds: Kamiokande II got 11, IMB got 8, Baksan got 5...but apparently Mount Blanc got none at this time. Then about 3 hours later, visible light arrived at earth indicating that the supernova had occurred. (Why didn't Mount Blanc receive any neutrinos when the other stations observed theirs and vice versa?)

The delay of the visible light (i.e., neutrinos traveling the 168,000 light-years faster than light by three hours) has been explained by arguing that the neutrinos (being the remarkably unsociable particles that they are) left the site of the star collapse immediately but the visible light (electromagnetic radiation) was tied up in the collapsing star for at least 3 hours.

Of course, that makes no sense. The star collapse itself involved 10^{46} joules worth of *matter* (like real protons, neutrons, electrons and stuff) and that matter could not move at the speed of light…much less all come together instantaneously (e.g., within 13 seconds). The neutrinos (if they exist) must have been released over a period of time comparable to the period of time that the flash of light from the star lasted. Both the electromagnetic waves (light) and the neutrinos (if they exist) are much faster than the movement of matter. Or are we to believe that the neutrinos themselves triggered the explosion?

And, where are the gamma rays from all the neutrino reactions that should have occurred inside the exploding star?

***

In September 2011 the CERN Laboratory in Switzerland was firing neutrinos 730 km through the earth to the INFN Gran Sasso Laboratory in Italy. Everything was carefully timed by atomic clocks and GPS positioning. The neutrinos were detected, but they arrived *before* they should have if they were traveling at the speed of light. In other words, we might have had to scrap the speed of light as an absolute in order to save the neutrino. The experiment was replicated several times and determined to have a high probability of reproducibility. A paper was published. Other labs jumped in. Scrapping the speed of light as an absolute was too much for the physicists who quickly found a number of sources of error in the experimental set up and expanded the uncertainty limits of the measurements to include the speed of light.

They then stopped.

www.ingramcontent.com/pod-product-compliance
Lightning Source LLC
Chambersburg PA
CBHW031925170526
45157CB00008B/3056